my life cycle

Robin

Published in the United States of America by Cherry Lake Publishing Group
Ann Arbor, Michigan
www.cherrylakepublishing.com

Reading Adviser: Marla Conn, MS, Ed., Literacy specialist, Read-Ability, Inc.
Book Designer: Jennifer Wahi
Illustrator: Jeff Bane

Photo Credits: © Brandon Blinkenberg/Shutterstock.com, 5; © Gabrielle Hovey/Shutterstock.com, 7; © Cheryl E. Davis/Shutterstock.com, 9; © Tony Campbell/Shutterstock.com, 11; © Jill Battaglia/Shutterstock.com, 13; © C. Hamilton/Shutterstock.com, 15; © JPBC/Shutterstock.com, 17; © Bonnie Taylor Barry/Shutterstock.com, 19; © Paul Reeves Photography/Shutterstock.com, 21; © Jeff Rzepka/Shutterstock.com, 23; Cover, 2-3, 10, 14, 22, 24, Jeff Bane

Cherry Lake Press is an imprint of Cherry Lake Publishing Group.

Library of Congress Cataloging-in-Publication Data

Names: Gray, Susan Heinrichs, author. | Bane, Jeff, 1957- illustrator.
Title: Robin / Susan H. Gray ; illustrated by Jeff Bane.
Description: Ann Arbor, Michigan : Cherry Lake Publishing, 2021. | Series: My life cycle | Includes index. | Audience: Grades 2-3
Identifiers: LCCN 2020030607 (print) | LCCN 2020030608 (ebook) | ISBN 9781534180048 (hardcover) | ISBN 9781534181755 (paperback) | ISBN 9781534181052 (pdf) | ISBN 9781534182769 (ebook)
Subjects: LCSH: Robins--Life cycles--Juvenile literature.
Classification: LCC QL696.P288 G73 2021 (print) | LCC QL696.P288 (ebook) | DDC 598.8/42--dc23
LC record available at https://lccn.loc.gov/2020030607
LC ebook record available at https://lccn.loc.gov/2020030608

Printed in the United States of America
Corporate Graphics

table of contents

About the author: Susan H. Gray has a master's degree in zoology. She loves writing science books, especially about animals. Susan lives in Arkansas with her husband, Michael. Every year, they get to watch baby robins grow up and leave the nest.

About the illustrator: Jeff Bane and his two business partners own a studio along the American River in Folsom, California, home of the 1849 Gold Rush. When Jeff's not sketching or illustrating for clients, he's either swimming or kayaking in the river to relax.

I'm a baby robin inside an egg. I share a nest with two other eggs.

Mom **snuggles** down and **incubates** us. Soon, I become the first to **hatch**.

At first, I’m a small and weak **nestling**. I have pink skin and patches of **wispy** feathers.

But what a big mouth I have! My parents feed me worms and bugs.

In time, my feathers grow and my muscles get stronger.

Sometimes, I stand up and **exercise** my wings. Finally, I am ready to fly.

I’m a **fledgling** now. I leap from the nest and flap as hard as I can. Uh-oh! I go down instead of up.

As I get older, I continue to practice. My dad watches me from the ground.

At 1 year old, I'm an adult. I can feed myself, and my flying has **improved**.

Now, it's time to find a **mate** and build a nest. I will be laying eggs soon!

glossary

exercise (EK-sur-size) to do an activity that strengthens the body

fledgling (FLEJ-ling) a young bird that is ready to fly

hatch (HACH) to break out of an egg

improved (im-PROOVD) to get better at doing something

incubates (ING-kyuh-bates) keeps something warm as it develops

mate (MATE) a partner to produce babies with

nestling (NEST-ling) a baby bird still in the nest

snuggles (SNUHG-uhlz) lies close to someone to guard and keep them warm

wispy (WISP-ee) thin and not very strong

index